초등
영어 읽기
1

초등 영어 읽기 ❶

초판발행 2020년 12월 21일

글쓴이 Contents Tree
그린이 정혜선, 조서아, 윤희재
엮은이 송지은, 진혜정, 김한나
기획 한동오
펴낸이 엄태상
영문감수 Kirsten March
디자인 진지화
오디오 전진우
마케팅 본부 이승욱, 전한나, 왕성석, 노원준, 조인선, 조성민
경영기획 마정인, 최성훈, 정다운, 김다미, 오희연
제작 전태준
물류 정종진, 윤덕현, 양희은, 신승진
펴낸곳 시소스터디
주소 서울시 종로구 자하문로 300 시사빌딩
주문 및 문의 1588-1582
팩스 02-3671-0510
홈페이지 www.sisostudy.com
이메일 sisostudy@sisadream.com
등록번호 제2019-000149호
ISBN 979-11-970830-8-2 63740

진 짜 진 짜

초등
영어 읽기

1

Introduction

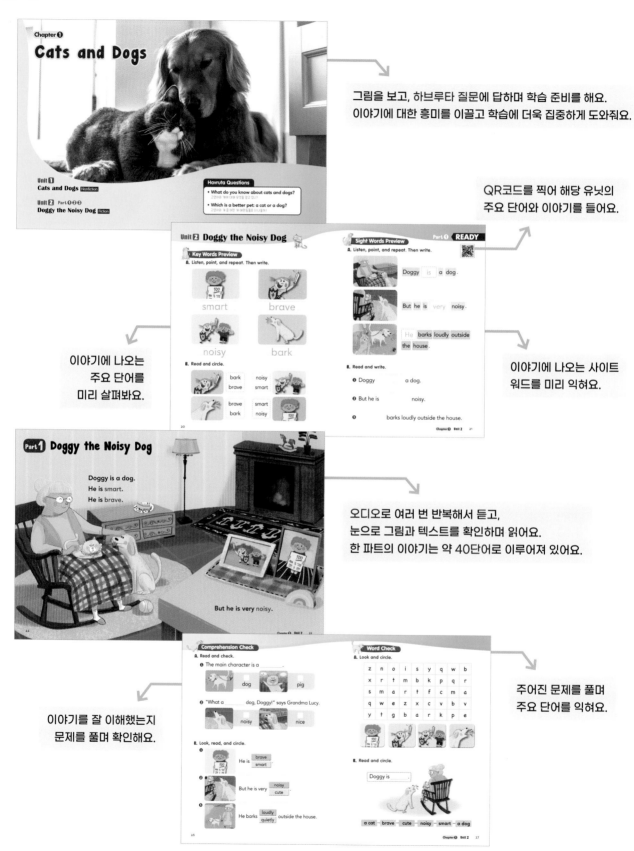

★ 〈진짜 진짜 초등 영어 읽기〉 이렇게 학습해 보세요.

그림을 보고, 하브루타 질문에 답하며 학습 준비를 해요.
이야기에 대한 흥미를 이끌고 학습에 더욱 집중하게 도와줘요.

QR코드를 찍어 해당 유닛의
주요 단어와 이야기를 들어요.

이야기에 나오는
주요 단어를
미리 살펴봐요.

이야기에 나오는 사이트
워드를 미리 익혀요.

오디오로 여러 번 반복해서 듣고,
눈으로 그림과 텍스트를 확인하며 읽어요.
한 파트의 이야기는 약 40단어로 이루어져 있어요.

이야기를 잘 이해했는지
문제를 풀며 확인해요.

주어진 문제를 풀며
주요 단어를 익혀요.

하브루타 가이드의 안내에 따라
워크북을 학습하며 **영어 실력** 뿐 아니라
생각하는 힘도 키워요!

알고 있는 단어와 모르는 단어를 체크
해요. 아는 것과 모르는 것을 정확히
파악하는 것만으로도 학습 효과는 눈에
띄게 달라요. (메타인지 효과)

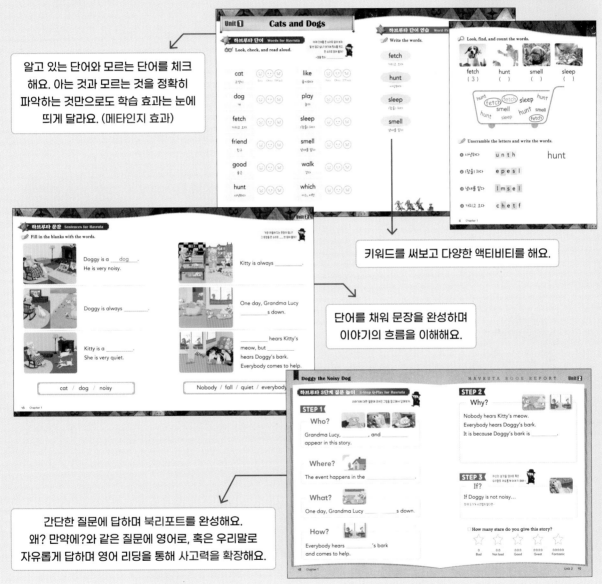

키워드를 써보고 다양한 액티비티를 해요.

단어를 채워 문장을 완성하며
이야기의 흐름을 이해해요.

간단한 질문에 답하며 북리포트를 완성해요.
왜? 만약에?와 같은 질문에 영어로, 혹은 우리말로
자유롭게 답하며 영어 리딩을 통해 사고력을 확장해요.

학습 안내 동영상을
확인하세요.

하브루타가 뭐예요?

하브루타란 서로 짝을 지어 질문하고 대답하면서 생각을 나누는 유대인의 전통 토론법이에요. 그 어느 누구와도 짝을 이룰 수 있겠지만, 학습 영역에서 그 짝은 함께 공부하는 친구들, 혹은 교사와 학생, 부모와 자녀로 이루어집니다.

 ## 하브루타 워크북 사용 설명서

1. 나의 하브루타 스터디 플랜: 오늘 학습할 내용을 확인해요.

2. 하브루타 단어: 단어를 큰 소리로 읽어 보며 쉬운 단어, 어려운 단어들을 체크해요.

3. 하브루타 단어 연습: 키워드를 손으로 여러 번 써보고, 연습 문제를 풀며 단어를 외워요.

4. **하브루타 3단계 질문 놀이:**

   Step 1. Who?(누가?), Where?(어디서?), What?(무엇을?), How?(어떻게?)에 답해요.

   Step 2. 읽은 내용을 바탕으로 Why?(왜?)에 답해요.

   Step 3. 새로운 가정에 대해 자유롭게 상상하면서 If?(만약에?)에 답해요.

각 코너마다 등장하는 하브루타 가이드의 질문에 답하며 학습을 진행하세요. 자연스럽게 하브루타식 워크북 활동을 할 수 있어요. 질문과 답이 꼭 영어일 필요는 없어요. 영어는 실력에 맞게 활용하되, 확장된 질문과 답을 자유롭게 하면 비판적으로 사고하는 능력을 키울 수 있어요.

하브루타 가이드

★★ 추천서 ★★

◇◇

『진짜 진짜 초등 영어 읽기』는 우리나라가 도입해야 할 하브루타식 질문이 자연스럽게 녹아 있는 탁월한 영어 리딩 학습서입니다. 하브루타라는 이름으로 창의적인 사고훈련과 공부법, 가족 대화법을 교육하는 전문가의 한사람으로서 우리 아이들에게 꼭 필요한 영어 학습서인 『진짜 진짜 초등 영어 읽기』를 추천합니다.

생각할 거리가 풍성하게 담겨있는 재미있는 이야기들은 단순한 패턴 반복식 영어 리딩의 한계를 극복하고 하브루타식 학습법으로 공부하기에 적합합니다. 이야기가 시작되기 전부터 흥미와 관심을 끄는 질문으로 아이들이 이야기에 집중하게 만들고, 단순히 학습의 흐름을 따라가는 것이 아니라 주도적으로 생각하며 이야기를 읽게 만듭니다. 또, 글을 읽은 후에는 사실 확인을 위한 질문 뿐 아니라 더 깊게 생각해 봐야할 질문을 통해 생각을 자극합니다. 아이들은 하브루타 학습법으로 공부하면서 영어 실력 뿐 아니라 생각하는 힘을 기를 수 있습니다.

워크북의 코너 곳곳에 마련된 질문들은 메타인지 효과로 학습자로 하여금 자신의 공부 모습을 들여다보게 합니다. 내가 알고 있는 것과 모르는 것을 정확히 아는 것만으로도 학습의 효과는 눈에 띄게 다릅니다. 생각을 자극하고 학습에 집중하게 하는 질문들을 읽고 답하는 것만으로도 하브루타 학습의 효과를 누릴 수 있습니다. 하브루타식 학습법이 접목된 『진짜 진짜 초등 영어 읽기』에서 창의적으로 생각하고 영어 리딩 실력을 높이시길 바랍니다.

교육학박사
하브루타창의인성교육연구소 소장   장성애

Contents

★ 한 챕터는 같은 소재로 쓰여진 Nonfiction과 Fiction 두 가지 글로 구성되어 있어요. ★

Chapter ❸ Vegetables

Chapter ❶
Cats and Dogs

학습 안내 동영상을
확인하세요.

Havruta Questions

- **What do you know about cats and dogs?**
  고양이와 개에 대해 무엇을 알고 있니?

- **Which is a better pet: a cat or a dog?**
  고양이와 개 중 어떤 게 애완동물로 더 나을까?

Unit **1** Cats and Dogs

A. Listen, point, and repeat. Then write.

sleep

fetch

hunt

smell

B. Read and match.

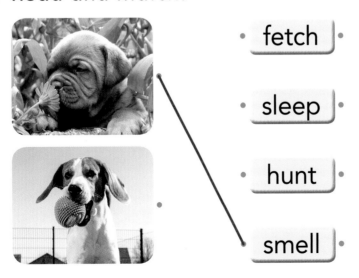

- fetch -

- sleep -

- hunt -

- smell -

Sight Words Preview

A. Listen, point, and repeat. Then write.

Cats **and** dogs are good friends.

Cats like *to* play.

Which do *you* like?

B. Read and write.

❶ Cats ___and___ dogs are good friends.

❷ Cats like _____ play.

❸ Which do _____ like?

Woof
Woof

Cats and Dogs

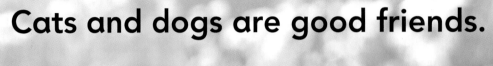

Cats and dogs are good friends.

Cats like to play.

Dogs like to walk.

Cats like to sleep.

Dogs like to fetch.

Cats like to hunt.

Dogs like to smell.

Which do you like?

Comprehension Check

A. Read and check.

1 This is a story about cats and _____.

 ✔ dogs    ☐ birds

2 Cats like to _____.

 ☐ walk    ☐ play

B. Look and check T(True) or F(False).

1

Dogs like to hunt.

T ☐   F ✔

2

Cats like to smell.

T ☐   F ☐

3

Dogs like to fetch.

T ☐   F ☐

18

Word Check

A. Complete the puzzle.

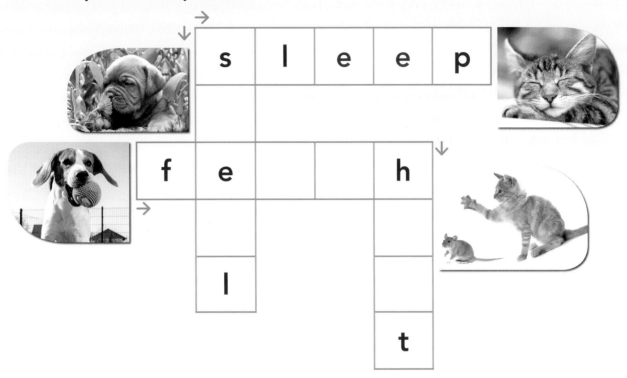

| | s | l | e | e | p |
| | | | | | |

| f | e | | | h |
| | | | | |

| | l | | | |

| | | | | t |

B. Look, read, and match.

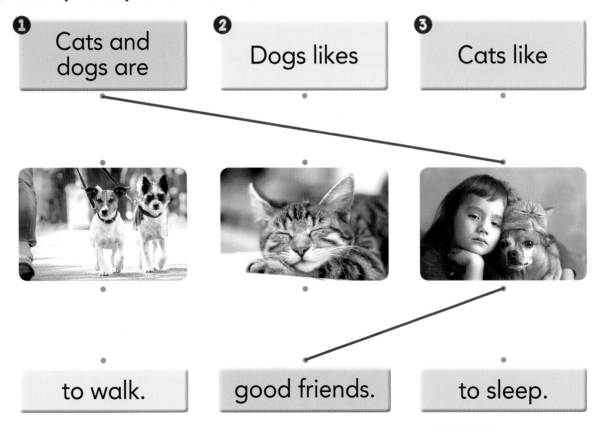

❶ Cats and dogs are

❷ Dogs likes

❸ Cats like

to walk.

good friends.

to sleep.

Doggy the Noisy Dog

Key Words Preview

A. Listen, point, and repeat. Then write.

smart

brave

noisy

bark

B. Read and circle.

bark	noisy
brave	smart

brave	smart
bark	noisy

Sight Words Preview

A. Listen, point, and repeat. Then write.

Doggy [is] a dog .

But he is [very] noisy .

[He] barks loudly outside the house .

B. Read and write.

❶ Doggy _____ a dog.

❷ But he is _____ noisy.

❸ _____ barks loudly outside the house.

Part 1 Doggy the Noisy Dog

Doggy is a dog.

He is smart.

He is brave.

But he is very noisy.

Doggy barks loudly inside the house.

He barks loudly outside the house.
"What a noisy dog, Doggy!"
says Grandma Lucy.

Comprehension Check

A. Read and check.

❶ The main character is a _____.

dog   pig

❷ "What a _____ dog, Doggy!" says Grandma Lucy.

noisy   nice

B. Look, read, and circle.

❶ He is
brave
smart
.

❷ But he is very
noisy
cute
.

❸ He barks
loudly
quietly
outside the house.

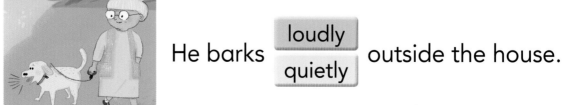

Word Check

A. Look and circle.

z	n	o	i	s	y	q	w	b
x	r	t	m	b	k	p	q	r
s	m	a	r	t	f	c	m	a
q	w	e	z	x	c	v	b	v
y	t	g	b	a	r	k	p	e

B. Read and circle.

Doggy is _____.

a cat — brave — cute — noisy — smart — a dog

Key Words Preview

A. Listen, point, and repeat. Then write.

cute

quiet

meow

house

B. Read and match.

· cute ·

· house ·

· meow ·

· quiet ·

Sight Words Preview

A. Listen, point, and repeat. Then write.

Kitty is a cat .

And she is very quiet .

She meows quietly outside the house .

B. Read and write.

❶ Kitty is _____ cat.

❷ And _____ is very quiet.

❸ She meows quietly outside _____ house.

Part 2 Doggy the Noisy Dog

Kitty is a cat.

She is smart.

She is cute.

And she is very quiet.

Kitty meows quietly inside the house.

She meows quietly outside the house.

"What a nice cat, Kitty!" says Grandma Lucy.

Comprehension Check

A. Read and check.

❶ Kitty is a _____ .

dog   cat

❷ "What a _____ cat, Kitty!" says Grandma Lucy.

nice   loud

B. Look, read, and circle.

❶

She is   brave / cute .

❷

And she is very   noisy / quite .

❸

She   barks / meows   quietly outside the house.

Word Check

A. Look and circle.

▶ x v c u t e q w r v

▶ p q u i e t b v c k

▶ g h x z q y m e o w

▶ h o u s e g v b q p

B. Read and circle.

Kitty is _____ .

a cat — nice — smart — cute — a dog — noisy

Unit 2 Doggy the Noisy Dog

Key Words Preview

A. Listen, point, and repeat. Then write.

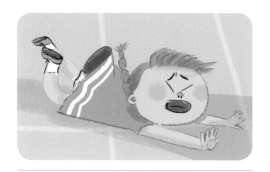

fall

move

hear

help

B. Read and circle.

fall
help

hear
fall

move
help

hear
move

36

Sight Words Preview

A. Listen, point, and repeat. Then write.

Grandma Lucy falls

down .

She can not move .

Kitty meows for help.

B. Read and write.

❶ Grandma Lucy falls _____ .

❷ She can _____ move.

❸ Kitty meows _____ help.

Part 3

Doggy the Noisy Dog

One day, Grandma Lucy falls down.

She can not move.

Kitty meows for help.

Nobody hears Kitty's meow.

Doggy barks for help.
Everybody hears Doggy's bark.

Everybody comes to help.

"Good boy, Doggy!" they say.

Comprehension Check

A. Read and check.

❶ One day, Grandma Lucy _____ down.

 ☐ falls    ☐ hears

❷ Everybody hears Doggy's _____.

 ☐ meow    ☐ bark

B. Look, read, and check.

❶

☐ Grandma Lucy can not help.

☐ Grandma Lucy can not move.

❷

☐ Kitty meows for help.

☐ Doggy meows for help.

❸

☐ Everybody comes to help.

☐ Nobody comes to help.

Word Check

A. Look and circle.

hear | fall | help

help | move | hear

fall | help | move

move | fall | hear

B. Look, read, and match.

1 Grandma Lucy

2 Nobody hears

3 Everybody hears

Kitty's meow.

Doggy's bark.

falls down.

Doggy the Noisy Dog

Doggy is a dog.

He is smart.

He is brave.

But he is very noisy.

Doggy barks loudly inside the house.

He barks loudly outside the house.

"What a noisy dog, Doggy!" says Grandma Lucy.

Kitty is a cat.

She is smart.

She is cute.

And she is very quiet.

Kitty meows quietly inside the house.

She meows quietly outside the house.

"What a nice cat, Kitty!" says Grandma Lucy.

One day, Grandma Lucy falls down.

She can not move.

Kitty meows for help.

Nobody hears Kitty's meow.

Doggy barks for help.

Everybody hears Doggy's bark.

Everybody comes to help.

"Good boy, Doggy!" they say.

Foxes

Havruta Questions

- **Look at the picture. What are they doing?**
  그림을 봐. 무엇을 하고 있는 걸까?

- **What is a fox like: clever or stupid?**
  여우는 어떤 거 같아? 영리할까? 어리석을까?

Unit 1 Foxes

Key Words Preview

A. Listen, point, and repeat. Then write.

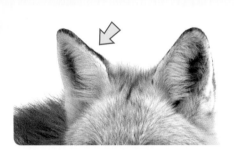

ear

tail

forest

desert

B. Read and match.

- forest
- ear
- tail
- desert

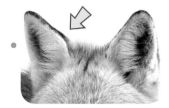

48

Sight Words Preview

READY

A. Listen, point, and repeat. Then write.

Foxes are wild animals .

They have large ears .

They live in deserts .

B. Read and write.

1 Foxes _____ wild animals.

2 They _____ large ears.

3 They live _____ deserts.

Foxes

Foxes are wild animals.

They have a bushy tail.

They have large ears.

Foxes live in many places.

They live in forests.

They live in deserts.

Foxes are good hunters.

They run fast.

They jump high.

Comprehension Check

A. Read and check.

❶ This is a story about _____.

 ☐ foxes    ☐ rabbits

❷ Foxes are good _____.

 ☐ places    ☐ hunters

B. Look and check T(True) or F(False).

❶

Foxes have small ears.

T ☐   F ☐

❷

Foxes live in deserts.

T ☐   F ☐

❸

Foxes run fast.

T ☐   F ☐

Word Check

A. Complete the puzzle.

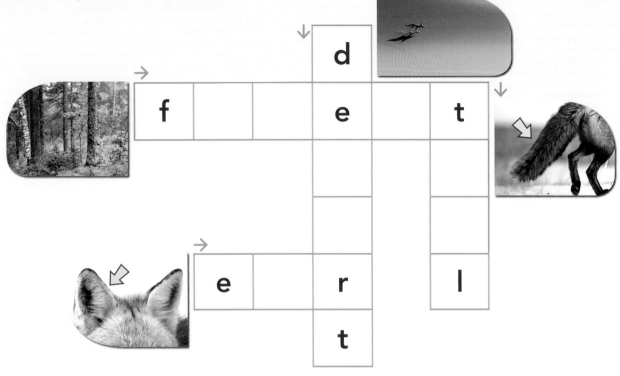

```
          d
  f       e       t
          r       l
  e       r
          t
```

B. Look, read, and match.

1 They live

2 They have

3 They jump

a bushy tail.

high.

in forests.

Unit 2 A Stupid Fox

 Key Words Preview

A. Listen, point, and repeat. Then write.

little

catch

dinner

dirty

B. Read and circle.

catch
little

dinner
catch

dinner
dirty

dirty
little

56

Sight Words Preview

A. Listen, point, and repeat. Then write.

They don't see the fox.

The fox catches one.

A dirty pig is not a tasty dinner.

B. Read and write.

❶ _____ don't see the fox.

❷ _____ fox catches one.

❸ A dirty pig _____ not a tasty dinner.

Part 1

A Stupid Fox

The three little pigs play in the woods.

They don't see the fox.

The fox catches one.
"Dinner time!" he says.

"He is dirty," say the two other pigs.
"A dirty pig is not a tasty dinner."

Comprehension Check

A. Read and check.

1 The main character is a _____.

 fox  dog

2 The three little _____ play in the woods.

 pigs  cats

B. Look, read, and circle.

1  They don't see the
rabbit
fox
.

2  "
Play
Dinner
time!" he says.

3  A
cute
dirty
pig is not a tasty dinner.

Word Check

A. Look and circle.

c	k	h	d	i	n	n	e	r
a	z	x	c	v	b	m	n	j
t	q	w	l	i	t	t	l	e
c	p	k	j	h	g	f	d	s
h	v	d	i	r	t	y	b	m

B. Look, read, and write.

_____ time!

A _____ pig is not a tasty dinner.

Unit 2 A Stupid Fox

Key Words Preview

A. Listen, point, and repeat. Then write.

want

tasty

wash

small

B. Read and match.

tasty

small

wash

want

64

Sight Words Preview

A. Listen, point, and repeat. Then write.

The fox wants a tasty dinner.

He washes the pig.

A small pig is not a tasty dinner.

B. Read and write.

❶ The fox wants _____ tasty dinner.

❷ _____ washes the pig.

❸ A small pig is _____ a tasty dinner.

Part 2

A Stupid Fox

The fox wants a tasty dinner.
He washes the pig.

"Dinner time!" says the fox.

"He is small," say the two other pigs.
"A small pig is not a tasty dinner."

Comprehension Check

A. Read and check.

1 The fox wants a _____ dinner.

 ☐ tasty    ☐ hungry

2 "He is _____," say the two other pigs.

 ☐ cute    ☐ small

B. Look, read, and circle.

1  He washes the [dog / pig] .

2  "Dinner time!" says the [fox / cat] .

3  A [nice / small] pig is not a tasty dinner.

Word Check

A. Look and circle.

 ▶ g f v d q **w a n t** z

 ▶ b p **t a s t y** v m x

 ▶ o f d z q c **w a s h**

 ▶ **s m a l l** v x p q w

B. Look, read, and write.

_____ time!

A _____ pig
is not a tasty dinner.

Unit 2 A Stupid Fox

Key Words Preview

A. Listen, point, and repeat. Then write.

feed

food

tired

asleep

B. Read and circle.

feed
asleep

tired
feed

tired
food

asleep
food

Sight Words Preview

A. Listen, point, and repeat. Then write.

He feeds the pig more and more food!

Now the fox feels tired.

Bedtime, you stupid fox!

B. Read and write.

❶ He feeds the pig more _____ more food!

❷ Now _____ fox feels tired.

❸ Bedtime, _____ stupid fox!

Part 3 A Stupid Fox

The fox wants a tasty dinner.

He feeds the pig more and more food!

Now the fox feels tired.

He falls asleep.

"Bedtime, you stupid fox!"
say the three little pigs.

Comprehension Check

A. Read and check.

❶ Now the fox feels _____.

 ☐ happy    ☐ tired

❷ "Bedtime, you stupid _____!" say the three little pigs.

 ☐ fox    ☐ cat

B. Look, read, and check.

☐ The pig wants a tasty dinner.
☐ The fox wants a tasty dinner.

☐ He feeds the pig more food.
☐ He feeds the dog more food.

☐ He falls asleep.
☐ He falls down.

Word Check

A. Look and circle.

tired | feed | asleep

food | asleep | tired

tasty | help | tired

food | asleep | feed

B. Look, read, and match.

1 The fox wants

2 He feeds the pig

3 Now the fox feels

tired.

a tasty dinner.

more food.

A Stupid Fox

The three little pigs play in the woods.

They don't see the fox.

The fox catches one.

"Dinner time!" he says.

"He is dirty," say the two other pigs.

"A dirty pig is not a tasty dinner."

The fox wants a tasty dinner.
He washes the pig.

"Dinner time!" says the fox.
"He is small," say the two other pigs.
"A small pig is not a tasty dinner."

The fox wants a tasty dinner.
He feeds the pig more and
more food!

Now the fox feels tired.
He falls asleep.
"Bedtime, you stupid fox!"
say the three little pigs.

Chapter ③
Vegetables

Havruta Questions

- **What kind of vegetables do you like to eat?**
  어떤 야채를 먹고 싶니?

- **Are there any vegetables you don't like? Why?**
  싫어하는 야채가 있니? 이유는?

Unit 1 Vegetables

Key Words Preview

A. Listen, point, and repeat. Then write.

vegetable

carrot

root

leaf

B. Read and match.

root

leaf

carrot

vegetable

84

Sight Words Preview

A. Listen, point, and repeat. Then write.

We eat some plants.

A carrot is a vegetable.

It is a fruit.

B. Read and write.

❶ We eat _____ plants.

❷ A carrot _____ a vegetable.

❸ _____ is a fruit.

Vegetables

We eat some plants.

They are vegetables.

A carrot is a vegetable.
It is a root.

A pumpkin is a vegetable.

It is a fruit.

Cabbage is a vegetable.

It is a leaf.

Which do you like to eat?

Comprehension Check

A. Read and check.

❶ This is a story about _____.

 □ vegetables  □ animals

❷ A carrot is a _____.

 □ fruit  □ root

B. Look and check T(True) or F(False).

  We eat some plants.

T □ F □

 A pumpkin is a root.

T □ F □

 Cabbage is a leaf.

T □ F □

Word Check

A. Complete the puzzle.

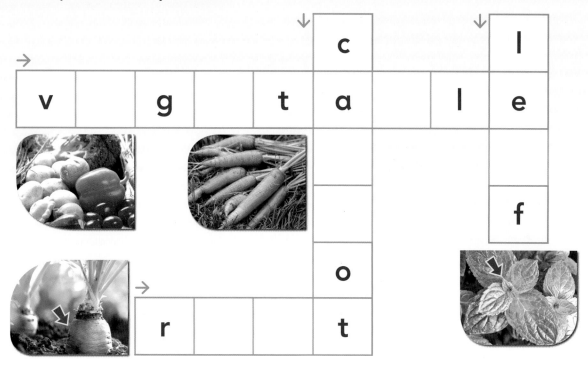

```
                          ↓
                        c              ↓
  →                                    l
  v       g       t    a        l      e

                                       f
                        o
      →
      r                 t
```

B. Look, read, and match.

❶ We eat   ❷ It is   ❸ Cabbage is

a root.   a leaf.   some plants.

Unit 2 The Great Turnip

Key Words Preview

A. Listen, point, and repeat. Then write.

farmer

turnip

wife

pull

B. Read and circle.

pull
wife

farmer
turnip

wife
turnip

farmer
pull

Sight Words Preview

A. Listen, point, and repeat. Then write.

It's a great turnip.

The farmer calls his wife.

They pull the turnip together.

B. Read and write.

❶ _____ a great turnip.

❷ The farmer calls _____ wife.

❸ _____ pull the turnip together.

Part 1 The Great Turnip

A farmer finds a big turnip.

"It's a big turnip."

The farmer calls his wife.

"It's a great turnip."
His wife calls their son.

"It's a big, great turnip."
They pull the turnip together.

Comprehension Check

A. Read and check.

❶ The main character is a _____.

 ☐ farmer    ☐ son

❷ They pull the _____ together.

 ☐ pumpkin    ☐ turnip

B. Look, read, and circle.

 ❶

It's a great cabbage / turnip .

❷

The farmer calls his wife / fruit .

❸

His wife calls their farmer / son .

Word Check

A. Look and circle.

z	f	a	r	m	e	r	x	p
w	q	b	v	z	w	t	c	k
i	a	z	p	u	l	l	v	m
f	p	d	g	v	w	q	b	x
e	k	y	t	u	r	n	i	p

B. Look, read, and write.

It's a _____ turnip.

It's a big _____.

It's a _____, _____ turnip.

Unit 2 The Great Turnip

Key Words Preview

A. Listen, point, and repeat. Then write.

king

queen

like

gold

B. Read and match.

- like
- king
- queen
- gold

Sight Words Preview

A. Listen, point, and repeat. Then write.

Let's take the turnip to the king .

The king will like it !

The king gives gold to the farmer .

B. Read and write.

❶ _____ take the turnip to the king.

❷ The king _____ like it!

❸ The king gives gold _____ the farmer.

Part 2 The Great Turnip

"Let's take the turnip to the king," says the farmer.

"The king will like it!"
says his wife.
"The queen will like it!"
says his son.

"I like it!"
says the king.

The king gives gold
to the farmer.

Comprehension Check

A. Read and check.

1 "Let's take the _____ to the king," says the farmer.

 ☐ turnip    ☐ pumpkin

2 "The _____ will like it!" says his wife.

 ☐ king    ☐ son

B. Look, read, and circle.

1

"The | king / queen | will like it!" says his son.

2

"I like it!" says the | farmer / king |.

3

The king gives | gold / turnip | to the farmer.

Word Check

A. Look and circle.

k i n g q w a n z t

b q u e e n v m x

h g f w u q l i k e

s q w z g o l d b v

B. Look, read and write.

Let's take the turnip to the _____.

The _____ will like it!

The _____ will like it!

Unit 2 The Great Turnip

Key Words Preview

A. Listen, point, and repeat. Then write.

rich

man

give

shout

B. Read and circle.

man
give

give
shout

rich
shout

man
rich

Sight Words Preview

A. Listen, point, and repeat. Then write.

Let's take gold to the king .

The king and the queen will like it !

What will I give you ?

B. Read and write.

❶ Let's _____ gold to the king.

❷ The king _____ the queen will like it.

❸ _____ will I give you?

Part 3 The Great Turnip

"Let's take gold to the king," says a rich man.

"The king and the queen will like it!"

"What will they give us?" says his wife.

"I like it!" says the king.

"What will I give you?"

"The great turnip!"
shouts the king.

Comprehension Check

A. Read and check.

1 "Let's take _____ to the king," says a rich man.

 ☐ gold    ☐ turnip

2 "The great _____!" shouts the king.

 ☐ gold    ☐ turnip

B. Look, read, and circle.

☐ "I like it!" says the king.

☐ "I like it!" says the rich man.

☐ What will I like you?

☐ What will I give you?

☐ "The great turnip!" shouts the king.

☐ "The great gold!" shouts the king.

Word Check

A. Look and circle.

king   rich   shout

rich   give   man

shout   like   rich

man   give   like

B. Look, read, and match.

❶ Let's take gold

❷ The king and the queen will

❸ What will I

like it!

to the king.

give you?

The Great Turnip

A farmer finds a big turnip.

"It's a big turnip."

The farmer calls his wife.

"It's a great turnip."

His wife calls their son.

"It's a big, great turnip."

They pull the turnip together.

"Let's take the turnip to the king,"
says the farmer.

"The king will like it!" says his wife.

"The queen will like it!" says his son.

"I like it!" says the king.
The king gives gold to the farmer.

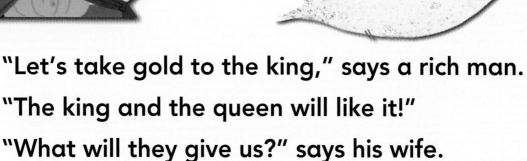

"Let's take gold to the king," says a rich man.

"The king and the queen will like it!"

"What will they give us?" says his wife.

"I like it!" says the king.

"What will I give you?"

"The great turnip!" shouts the king.

My Word List

asleep 잠이 든

He falls _____. (p.76)

bark 짖다

Doggy _____s loudly inside the house. (p.24)

brave 용감한

He is _____. (p.22)

carrot 당근

A _____ is a vegetable. (p.87)

catch 붙잡다

The fox _____es one. (p.59)

cute 귀여운

She is _____. (p.30)

d

desert 사막

They live in _____s. (p.52)

dinner 저녁 식사

"_____ time!" he says. (p.59)

dirty 더러운

"He is _____," say the two other pigs. (p.60)

e

ear 귀

They have large _____s. (p.51)

f

fall 넘어지다, 쓰러지다

One day, Grandma Lucy _____s down. (p.38)

farmer 농부

A _____ finds a big turnip. (p.94)

feed 밥을 먹이다

He _____s the pig. (p.75)

fetch 가지고 오다

Dogs like to _____. (p.16)

food 음식

More and more _____! (p.75)

forest 숲

They live in _____s. (p.52)

g

give 주다

"What will they _____ us?"
says his wife. (p.111)

gold 금

The king gives _____ to the
farmer. (p.105)

h

hear 듣다

Nobody _____s Kitty's meow.
(p.39)

help 도움

Kitty meows for _____. (p.39)

house 집

Kitty meows quietly inside the

_____. (p.32)

hunt 사냥하다

Cats like to _____. (p.17)

k

king 왕

"Let's take the turnip to the

_____," says the farmer. (p.102)

l

leaf (나뭇)잎

It is a _____. (p.89)

like 좋아하다

"The king will _____ it!" says his wife. (p.103)

little 어린

The three _____ pigs play in the woods. (p.58)

man (성인) 남자

"Let's take gold to the king," says a rich _____. (p.110)

meow 야옹하고 울다

Kitty _____s quietly inside the house. (p.32)

move 움직이다

She can not _____. (p.38)

noisy 시끄러운

But he is very _____. (p.23)

p

pull 뽑다, 당기다

They _____ the turnip together. (p.97)

q

queen 여왕

"The _____ will like it!" says his son. (p.103)

quiet 조용한

And she is very _____. (p.30)

r

rich 부유한

"Let's take gold to the king," says a _____ man. (p.110)

root 뿌리

It is a _____. (p.87)

s

shout 외치다

"The great turnip!" _____s the king. (p.113)

sleep (잠을) 자다

Cats like to _____. (p.16)

small 작은

"He is _____," say the two other pigs. (p.69)

smart 영리한, 똑똑한

He is _____. (p.22)

smell 냄새를 맡다

Dogs like to _____. (p.17)

t

tail 꼬리

They have a bushy _____. (p.51)

tasty 맛있는

The fox wants a _____ dinner. (p.66)

tired 피곤한, 지친

Now the fox feels _____. (p.76)

turnip 순무

"It's a big _____." (p.94)

vegetable 채소, 야채

They are _____s. (p.86)

want 원하다

The fox _____s a tasty dinner.
(p.66)

wash 씻다

He _____es the pig. (p.66)

wife 아내

The farmer calls his _____. (p.94)

진짜진짜

초등
영어 읽기

하브루타 워크북

1

SISO Study

진짜진짜
초등 영어 읽기
하브루타 워크북
1

Contents

하브루타 워크북 이렇게 학습하세요.

학습 안내 동영상을
확인하세요.

❶ 먼저 스토리북을 학습하고 와요.

❷ 하브루타 워크북을 펴고 학습한 스토리북에 해당하는 **하브루타 단어** 와 **하브루타 단어 연습** 을 해요.

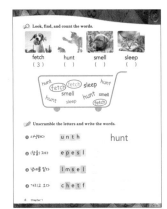

❸ 스토리 하나가 끝나면 **하브루타 문장** 에서 이야기에 대한 내용을 정리하고, **하브루타 3단계 질문 놀이** 를 통해 북리포트는 작성합니다. **LET'S THINK!** 에서는 자유롭게 질문하고 답해요.

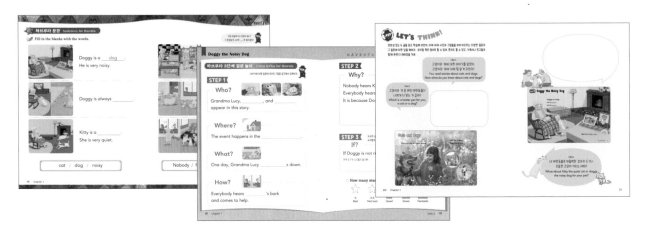

★ My Havruta Study Plan ★ 학습한 부분을 체크하고 날짜도 기록해 보세요.

	스토리북	워크북	학습한 날짜
Chapter ❶ Unit ❶ **Cats and Dogs**	10~19p ☐	4~6p ☐	____ . ____
Chapter ❶ Unit ❷ Part 1 **Doggy the Noisy Dog**	20~27p ☐	7~9p ☐	____ . ____
Chapter ❶ Unit ❷ Part 2 **Doggy the Noisy Dog**	28~35p ☐	10~12p ☐	____ . ____
Chapter ❶ Unit ❷ Part 3 **Doggy the Noisy Dog**	36~45p ☐	13~21p ☐	____ . ____
Chapter ❷ Unit ❶ **Foxes**	46~55p ☐	22~24p ☐	____ . ____
Chapter ❷ Unit ❷ Part 1 **A Stupid Fox**	56~63p ☐	25~27p ☐	____ . ____
Chapter ❷ Unit ❷ Part 2 **A Stupid Fox**	64~71p ☐	28~30p ☐	____ . ____
Chapter ❷ Unit ❷ Part 3 **A Stupid Fox**	72~81p ☐	31~39p ☐	____ . ____
Chapter ❸ Unit ❶ **Vegetables**	82~91p ☐	40~42p ☐	____ . ____
Chapter ❸ Unit ❷ Part 1 **The Great Turnip**	92~99p ☐	43~45p ☐	____ . ____
Chapter ❸ Unit ❷ Part 2 **The Great Turnip**	100~107p ☐	46~48p ☐	____ . ____
Chapter ❸ Unit ❷ Part 3 **The Great Turnip**	108~117p ☐	49~57p ☐	____ . ____

Unit **1** Cats and Dogs

아래 단어를 큰 소리로 읽어 보자.
몇 번 읽고 싶니? 여기에 횟수를 적고,
큰 소리로 읽어 볼까?

• 읽을 횟수: _____

Look, check, and read aloud.

cat 고양이	Easy Okay Difficult	like 좋아하다	Easy Okay Difficult
dog 개		play 놀다	
fetch 가지고 오다		sleep (잠을) 자다	
friend 친구		smell 냄새를 맡다	
good 좋은		walk 걷다	
hunt 사냥하다		which 어느, 어떤	

Write the words.

fetch

가지고 오다

hunt

사냥하다

sleep

(잠을) 자다

smell

냄새를 맡다

외운 단어를 모두 적어보고, 힘들지 않게
잘 외워진 단어는 무엇인지 말해 볼까?

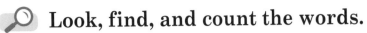 **Look, find, and count the words.**

fetch   hunt   smell   sleep

(3)   (   )   (   )   (   )

✏️ **Unscramble the letters and write the words.**

1 사냥하다   u n t h   hunt

2 (잠을) 자다   e p e s l

3 냄새를 맡다   l m s e l

4 가지고 오다   c h e t f

Doggy the Noisy Dog

하브루타 단어 Words for Havruta

아래 단어를 큰 소리로 읽어 보자.
몇 번 읽고 싶니? 여기에 횟수를 적고,
큰 소리로 읽어 볼까?

• 읽을 횟수: _____

Look, check, and read aloud.

bark 짖다	 Easy Okay Difficult	**noisy** 시끄러운
brave 용감한		**outside** 밖에서
grandma 할머니		**say** 말하다
house 집		**smart** 영리한, 똑똑한
inside 안에서		**very** 매우
loudly 큰 소리로, 시끄럽게		**what** 정말, 얼마나

noisy 쪽 얼굴 이미지: Easy   Okay   Difficult
outside:
say:
smart:
very:
what:

✏️ Write the words.

bark

짓다

brave

용감한

noisy

시끄러운

smart

똑똑한, 영리한

외우기 어려운 단어를 순서대로 3개만 써 봐.
왜 안 외워질까?

✏️ **Write the words correctly.**

bark

bark

noisy

brave

smart

🔍 **Solve the puzzle.**

n	**o**	**i**	**s**	**y**	

b

① He is very

n<u>oisy</u>____.

② He is b_____.

③ Doggy b_____s
loudly inside
the house.

④ He is s_____.

하브루타 단어 Words for Havruta

 아래 단어를 큰 소리로 읽어 보자.
몇 번 읽고 싶니? 여기에 횟수를 적고,
큰 소리로 읽어 볼까?

• 읽을 횟수: _____

Look, check, and read aloud.

cute		quiet	
귀여운	Easy Okay Difficult	조용한	Easy Okay Difficult
house		quietly	
집		조용히	
inside		say	
안에서		말하다	
meow		smart	
야옹하고 울다		영리한, 똑똑한	
nice		very	
좋은		매우	
outside		what	
밖에서		정말, 얼마나	

Write the words.

cute
귀여운

house
집

meow
야옹하고 울다

quiet
조용한

'나만의 단어 외우는 방법'은 어떤 게 있을까?

Match the letters to the words.

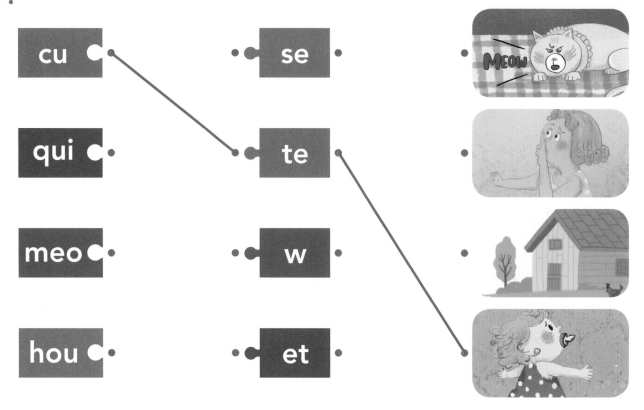

cu · · se

qui · · te

meo · · w

hou · · et

Cross out the wrong ones.

house   house   house   ~~huose~~

moew   meow   meow   meow

cute   cute   tuce   cute

quiet   qiuet   quiet   quiet

Doggy the Noisy Dog

 하브루타 단어 **Words for Havruta**

 Look, check, and read aloud.

 아래 단어를 큰 소리로 읽어 보자.
몇 번 읽고 싶니? 여기에 횟수를 적고,
큰 소리로 읽어 볼까?
• 읽을 횟수: _____

bark 짖다	Easy Okay Difficult
hear 듣다	Easy Okay Difficult
can not ~할 수 없다	
help 도움	
come 오다	
move 움직이다	
everybody 모두	
nobody 아무도…않다	
fall 넘어지다, 쓰러지다	
one day 어느 날	
good 잘 한, 훌륭한	

Write the words.

fall
넘어지다, 쓰러지다

hear
듣다

help
도움

move
움직이다

가장 좋아하는 단어를 쓰고,
빈 종이에 그림으로 그려보자.

✏️ **Connect and write the words.**

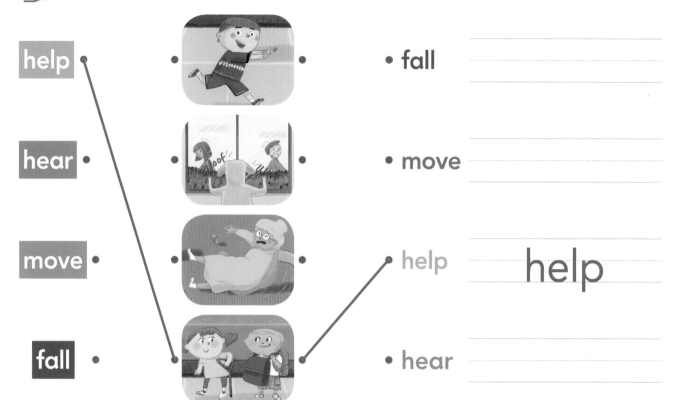

help   • fall

hear   • move

move   • help   help

fall   • hear

✏️ **Cross out the wrong letters and write the words.**

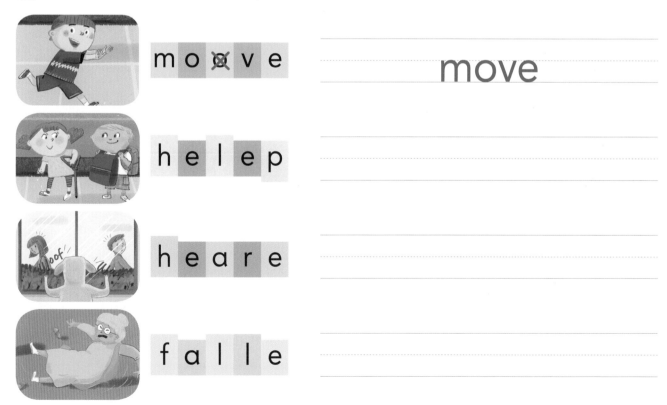

m o ⊗ v e   move

h e l e p

h e a r e

f a l l e

Fill in the blanks with the words.

Doggy is a ___dog___.
He is very noisy.

Doggy is always _____.

Kitty is a _____.
She is very quiet.

cat / dog / noisy

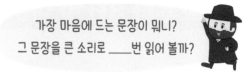

Kitty is always _____.

One day, Grandma Lucy _____s down.

_____ hears Kitty's meow, but _____ hears Doggy's bark. Everybody comes to help.

Nobody / fall / quiet / everybody

Doggy the Noisy Dog

이야기에 대한 질문에 주어진 그림을 참고해서 답해보자.

STEP 1

Who?

Grandma Lucy, _____, and _____ appear in this story.

Where?

The event happens in the _____.

What?

One day, Grandma Lucy _____s down.

How?

Everybody hears _____'s bark and comes to help.

STEP 2
Why?

Nobody hears Kitty's meow.
Everybody hears Doggy's bark.
It is because Doggy's bark is _____.

STEP 3

자신의 생각을 영어로 혹은
우리말로 자유롭게 이야기 해봐~

If?

If Doggy is not noisy...

만약 도기가 시끄럽지 않다면…

☐ **How many stars do you give this story?**

    ☆

| ☆ | ☆☆ | ☆☆☆ | ☆☆☆☆ | ☆☆☆☆☆ |
| Bad | Not bad | Good | Great | Fantastic |

LET'S THINK!

연관성 있는 두 글을 읽고 학습해 보았어. 이제 아래 사진과 그림들을 보며 떠오르는 다양한 질문과 그 질문에 대한 답을 해보자. 우리말 혹은 영어로 할 수 있어. 혼자도 할 수 있고, 가족이나 친구들과 함께 하면 더 재미있을 거야.

(예시)
고양이와 개에 대한 이야기를 읽었어.
고양이와 개에 대해 뭘 알게 되었어?
You read stories about cats and dogs.
Now what do you know about cats and dogs?

(예시)
고양이와 개 중 어떤 애완동물이
너에게 더 맞는 거 같아?
Which is a better pet for you:
a cat or a dog?

(예시)
네 애완동물로 떠들썩한 강아지 도기나
조용한 고양이 키티는 어때?

What about Kitty the quiet cat or Doggy
the noisy dog for your pet?

하브루타 단어   Words for Havruta

 Look, check, and read aloud.

아래 단어를 큰 소리로 읽어 보자.
몇 번 읽고 싶니? 여기에 횟수를 적고,
큰 소리로 읽어 볼까?

• 읽을 횟수: _____

animal 동물	Easy	Okay	Difficult
bushy 털이 복슬복슬한			
desert 사막			
ear 귀			
fast 빠른			
forest 숲			
fox 여우			

good 훌륭한, 뛰어난	Easy	Okay	Difficult
have 가지고 있다			
high 높이			
hunter 사냥꾼			
jump 점프하다			
large 큰			
live 살다			

many
많은
Easy Okay Difficult

place
장소

run
달리다

tail
꼬리
Easy Okay Difficult

wild
야생의

하브루타 단어 연습 Word Practice for Havruta

Write the words.

desert
사막

ear
귀

forest
숲

tail
꼬리

 Look, find, and count the words.

desert   ear   forest   tail

(   )   (   )   (   )   (   )

ear   tail   ear   tail
tail   desert   desert
forest   ear   tail
ear

Unscramble the letters and write the words.

❶ 숲   e o s r f t   _____

❷ 꼬리   i l t a   _____

❸ 사막   e e r s d t   _____

❹ 귀   a r e   _____

하브루타 단어 Words for Havruta

Look, check, and read aloud.

아래 단어를 큰 소리로 읽어 보자.
몇 번 읽고 싶니? 여기에 횟수를 적고,
큰 소리로 읽어 볼까?

• 읽을 횟수: _____

	Easy	Okay	Difficult		Easy	Okay	Difficult
three 3, 셋				**pig** 돼지			
catch 붙잡다				**see** 보다			
dinner 저녁 식사				**play** 놀다			
dirty 더러운				**tasty** 맛있는			
little 어린				**time** 시간			
not …아니다				**two** 2, 둘			
other 다른				**woods** 숲			

✏️ **Write the words.**

catch
붙잡다

dinner
저녁 식사

dirty
더러운

little
어린

외운 단어를 모두 적어보고, 힘들지 않게
잘 외워진 단어는 무엇인지 말해 볼까?

✎ Write the words correctly.

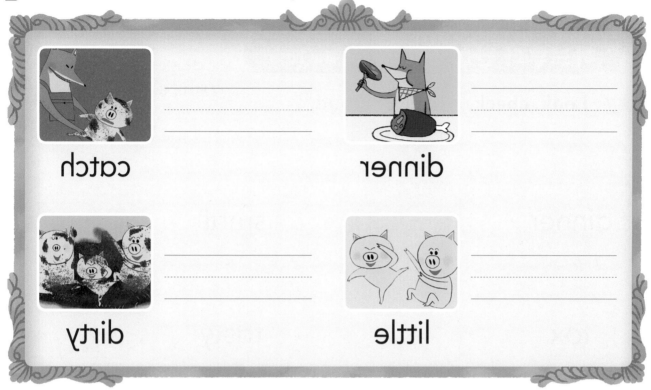

catch

dinner

dirty

little

🔍 Solve the puzzle.

❹↓

c

❸↓

d

❶→ l

❷→ d

❶ The three l_____ pigs play in the woods.

❷ "He is d_____," say the two other pigs.

❸ "D_____ time!" he says.

❹ The fox c_____es one.

하브루타 단어   Words for Havruta

Look, check, and read aloud.

아래 단어를 큰 소리로 읽어 보자.
몇 번 읽고 싶니? 여기에 횟수를 적고,
큰 소리로 읽어 볼까?
• 읽을 횟수: _____

dinner	Easy	Okay	Difficult	small	Easy	Okay	Difficult
저녁 식사				작은			
fox				tasty			
여우				맛있는			
not				time			
…아니다				시간			
other				two			
다른				2, 둘			
pig				want			
돼지				원하다			
say				wash			
말하다				씻다			

하브루타 단어 연습 Word Practice for Havruta

 Write the words.

small
작은

tasty
맛있는

want
원하다

wash
씻다

외우기 어려운 단어를 순서대로 3개만 써 봐.
왜 안 외워질까?

Match the letters to the words.

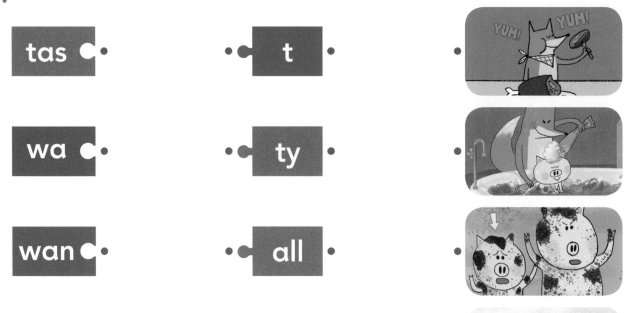

tas | t
wa | ty
wan | all
sm | sh

Cross out the wrong ones.

small	small	small	msall
tasty	tatsy	tasty	tasty
awnt	want	want	want
wash	wash	wash	sawh

A Stupid Fox

 하브루타 단어 Words for Havruta

 아래 단어를 큰 소리로 읽어 보자.
몇 번 읽고 싶니? 여기에 횟수를 적고,
큰 소리로 읽어 볼까?

• 읽을 횟수: _____

👓 **Look, check, and read aloud.**

feed 밥을 먹이다	Easy Okay Difficult	**more and more** 점점 더 많은
asleep 잠이 든		**now** 이제
bedtime 취침 시간		**stupid** 멍청한, 바보 같은
fall 떨어지다, 빠지다		**three** 3, 셋
feel 느끼다		**tired** 피곤한, 지친
food 음식		**want** 원하다
little 어린		

Write the words.

feed
밥을 먹이다

_____

asleep
잠이 든

_____

food
음식

_____

tired
피곤한, 지친

_____

'나만의 단어 외우는 방법'은 어떤 게 있을까?

_____

_____

_____

✏️ **Connect and write the words.**

feed •

asleep •

food •

tired •

• tired

• food

• feed

• asleep

_____

_____

_____

_____

✏️ **Cross out the wrong letters and write the words.**

 a s i l e e p

 f e a e d

 t a i r e d

 f o o d e

Fill in the blanks with the words.

The three little pigs play in the woods.
The fox _____es one.

The pig is _____.

The fox _____es the pig.

dirty / catch / wash

○○에게 들려주고 싶은 문장이 있니?
그 문장을 외워서 ○○ 앞에서
큰 소리로 말해 보자.

The pig is _____.

The fox _____s the pig.

The fox feels _____
and falls _____.

asleep / feed / tired / small

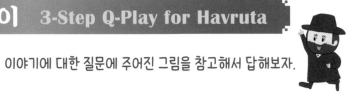

이야기에 대한 질문에 주어진 그림을 참고해서 답해보자.

STEP 1

Who?

Three little _____s and a _____ appear in this story.

Where?

The event happens in the _____ and the _____'s house.

What?

The fox _____es the pig. He wants it for his _____.

How?

The fox washes the pig. He feeds the pig. He feels _____ and falls _____.

STEP 2

Why?

The fox washes and feeds the pig.

It is because the fox is _____.

STEP 3

자신의 생각을 영어로 혹은
우리말로 자유롭게 이야기 해봐~

If?

If the fox doesn't wash or feed the pig...

여우가 돼지를 씻기거나 먹이지 않는다면…

☐ **How many stars do you give this story?**

☆	☆☆	☆☆☆	☆☆☆☆	☆☆☆☆☆
Bad	Not bad	Good	Great	Fantastic

연관성 있는 두 글을 읽고 학습해 보았어. 이제 아래 사진과 그림들을 보며 떠오르는 다양한 질문과 그 질문에 대한 답을 해보자. 우리말 혹은 영어로 할 수 있어. 혼자도 할 수 있고, 가족이나 친구들과 함께 하면 더 재미있을 거야.

(예시)
여우에 대한 두 글을 읽었는데,
여우에 대해 뭘 알게 되었어?

You read stories about foxes.
Now what do you know about foxes?

Foxes

Foxes are wild animals.
They have a bushy tail.
They have large ears.

Chapter 2 Unit 1   51

50

Part 1 **A Stupid Fox**

The three little pigs play in the woods.
They don't see the fox.

The fox catches one.
"Dinner time!" he says.

58

Chapter 2 Unit 2 59

(예시)

여우는 영리하다고 생각해?
아니면 어리석다고 생각해? 왜 그렇게 생각해?
Do you think they are clever or stupid?
Why?

Vegetables

하브루타 단어 **Words for Havruta**

아래 단어를 큰 소리로 읽어 보자.
몇 번 읽고 싶니? 여기에 횟수를 적고,
큰 소리로 읽어 볼까?

• 읽을 횟수: _____

Look, check, and read aloud.

vegetable
채소, 야채

Easy Okay Difficult

plant
식물

Easy Okay Difficult

cabbage
양배추

pumpkin
호박

carrot
당근

root
뿌리

eat
먹다

some
일부의, 어떤

fruit
과일

which
어느, 어떤

leaf
(나뭇)잎

하브루타 단어 연습 Word Practice for Havruta

 Write the words.

vegetable

채소, 야채

carrot

당근

leaf

(나뭇)잎

root

뿌리

가장 좋아하는 단어를 쓰고,
빈 종이에 그림으로 그려보자.

 Look, find, and count the words.

vegetable   carrot   leaf   root
( )   ( )   ( )   ( )

root carrot vegetable
leaf vegetable carrot
root leaf
carrot leaf root carrot

Unscramble the letters and write the words.

❶ (나뭇)잎   a e f l

❷ 당근   r r o a c t

❸ 뿌리   o t r o

❹ 채소, 야채   a e b e l e v t g

하브루타 단어 Words for Havruta

 Look, check, and read aloud.

아래 단어를 큰 소리로 읽어 보자.
몇 번 읽고 싶니? 여기에 횟수를 적고,
큰 소리로 읽어 볼까?

• 읽을 횟수: _____

big 큰  Easy Okay Difficult	**pull** 당기다, 뽑다  Easy Okay Difficult
call 부르다 	**son** 아들 
farmer 농부 	**together** 함께 
find 찾다, 발견하다 	**turnip** 순무 
great 큰, 엄청난 	**wife** 아내 

Write the words.

farmer
농부

pull
당기다, 뽑다

turnip
순무

wife
아내

외운 단어를 모두 적어보고, 힘들지 않게
잘 외워진 단어는 무엇인지 말해 볼까?

✏️ **Write the words correctly.**

farmer

pull

turnip

wife

🔍 **Solve the puzzle.**

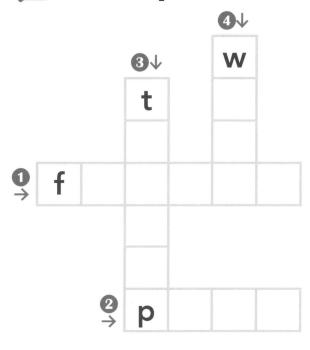

❶ A f_____ finds a big turnip.

❷ They p_____ the turnip together.

❸ "It's a big t_____."

❹ The farmer calls his w_____.

The Great Turnip

하브루타 단어   Words for Havruta

아래 단어를 큰 소리로 읽어 보자.
몇 번 읽고 싶니? 여기에 횟수를 적고,
큰 소리로 읽어 볼까?

• 읽을 횟수: _____

Look, check, and read aloud.

farmer	Easy Okay Difficult	queen	Easy Okay Difficult
농부		여왕	
give		son	
주다		아들	
gold		take	
금		가지고 가다	
king		turnip	
왕		순무	
let's		will	
~하자		...할 것이다	
like			
좋아하다			

하브루타 단어 연습   Word Practice for Havruta

✏ **Write the words.**

gold
금

king
왕

like
좋아하다

queen
여왕

외우기 어려운 단어를 순서대로 3개만 써 봐.
왜 안 외워질까?

Match the letters to the words.

qu	ng	
ki	ke	
li	een	
go	ld	

Cross out the wrong ones.

like   like   like   lkie

qeeun   queen   queen   queen

gold   godl   gold   gold

king   king   gink   king

The Great Turnip

하브루타 단어   **Words for Havruta**

 Look, check, and read aloud.

아래 단어를 큰 소리로 읽어 보자.
몇 번 읽고 싶니? 여기에 횟수를 적고,
큰 소리로 읽어 볼까?

• 읽을 횟수: _____

give				rich			
주다	Easy	Okay	Difficult	부유한	Easy	Okay	Difficult

king				shout			
왕				외치다			

let's				take			
~하자				가지고 가다			

like				what			
좋아하다				무엇, 어떤			

man				wife			
(성인) 남자				아내			

queen				will			
여왕				…할 것이다			

✏️ **Write the words.**

give
주다

man
(성인) 남자

rich
부유한

shout
외치다

'나만의 단어 외우는 방법'은 어떤 게 있을까?

✏️ **Connect and write the words.**

give •

• give _____

rich •

• man _____

shout •

• shout _____

man •

• rich _____

✏️ **Cross out the wrong letters and write the words.**

   m e a n   _____

   s c h o u t   _____

   r i e c h   _____

   g i e v e   _____

Fill in the blanks with the words.

A _____ finds a big turnip.

The farmer, his wife and his son _____ the turnip together.

The farmer says, "Let's take the turnip to the _____."

pull / farmer / king

가장 마음에 드는 문장이 뭐니?
그 문장을 큰 소리로 ____번 읽어 볼까?

The king likes the turnip, and gives _____ to the farmer.

A _____ man says, "Let's take gold to the king."

The king likes the gold, and gives the great _____ to the rich man.

gold / turnip / rich

STEP 1

이야기에 대한 질문에 주어진 그림을 참고해서 답해보자.

Who?

A _____, a _____ man, and
a _____ appear in this story.

Where?

The event happens in the _____'s field
and the _____'s palace.

What?

The farmer takes the big _____ to the king.
The rich man takes _____ to the king.

How?

The king gives _____ to the farmer
and the great _____ to the rich man.

*field 밭 palace 궁전

STEP 2

Why?

The farmer takes the turnip to the king.

It is because he thinks the king _____

_____.

The rich man takes gold to the king.

It is because he thinks the king _____

_____.

STEP 3

자신의 생각을 영어로 혹은
우리말로 자유롭게 이야기 해봐~

If?

If the farmer finds gold in his field...

농부가 밭에서 황금을 발견한다면…

☐ **How many stars do you give this story?**

☆	☆☆	☆☆☆	☆☆☆☆	☆☆☆☆☆
Bad	Not bad	Good	Great	Fantastic

LET'S THINK!

연관성 있는 두 글을 읽고 학습해 보았어. 이제 아래 사진과 그림들을 보며 떠오르는 다양한 질문과 그 질문에 대한 답을 해보자. 우리말 혹은 영어로 할 수 있어. 혼자도 할 수 있고, 가족이나 친구들과 함께 하면 더 재미있을 거야.

(예시)
야채를 키워 본 적이 있니?
어떤 야채였어?

Have you ever planted vegetables?
What were they?

Part 1 **The Great Turnip**

A farmer finds a big turnip.
"It's a big turnip."
The farmer calls his wife.

94   Chapter 3 Unit 2   95

(예시)

어떤 야채를 키워보고 싶어?

이유는 뭐야?

What vegetables do you want to plant?
Why?

57

Storybook 1 정답

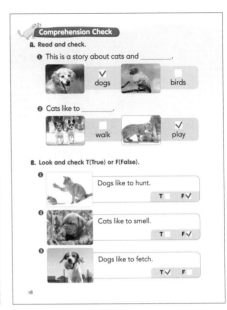

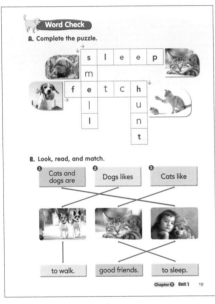

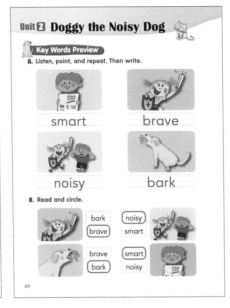

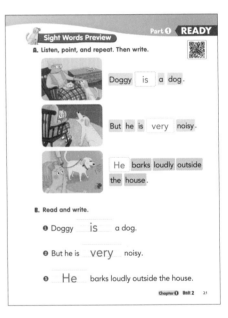

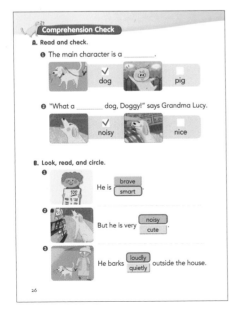

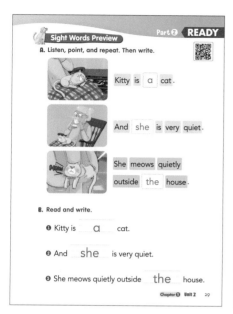

Sight Words Preview — Part 2 · READY

A. Listen, point, and repeat. Then write.

Kitty is [a] cat.

And [she] is very [quiet].

She [meows] [quietly] outside [the] [house].

B. Read and write.

❶ Kitty is **a** cat.

❷ And **she** is very quiet.

❸ She meows quietly outside **the** house.

Chapter 1 Unit 2 29

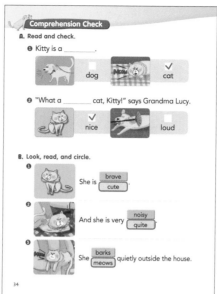

Comprehension Check

A. Read and check.

❶ Kitty is a _____.
dog ☐ / cat ✔

❷ "What a _____ cat, Kitty!" says Grandma Lucy.
nice ✔ / loud ☐

B. Look, read, and circle.

❶ She is [brave / **cute**]

❷ And she is very [**noisy** / quite]

❸ She [barks / **meows**] quietly outside the house.

34

Word Check

A. Look and circle.

x v (c u t e) q w r v

p (q u i e t) b v c k

g h x z q y (m e o w)

(h o u s e) g v b q p

B. Read and circle.

Kitty is _____.

a cat — nice — smart — **cute** — a dog — noisy

Chapter 1 Unit 2 35

Unit 2 Doggy the Noisy Dog

Key Words Preview

A. Listen, point, and repeat. Then write.

fall

move

hear

help

B. Read and circle.

[**fall** / help]   [hear / **fall**]

[**move** / help]   [**hear** / move]

36

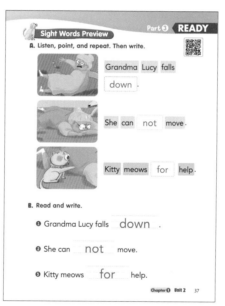

Sight Words Preview — Part 3 · READY

A. Listen, point, and repeat. Then write.

Grandma Lucy falls [down].

She [can] [not] [move].

Kitty [meows] [for] [help].

B. Read and write.

❶ Grandma Lucy falls **down**.

❷ She can **not** move.

❸ Kitty meows **for** help.

Chapter 1 Unit 2 37

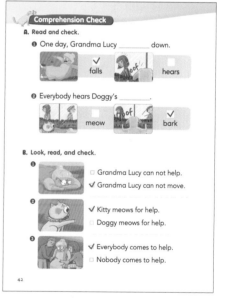

Comprehension Check

A. Read and check.

❶ One day, Grandma Lucy _____ down.
falls ✔ / hears ☐

❷ Everybody hears Doggy's _____.
meow ☐ / bark ✔

B. Look, read, and check.

❶ ☐ Grandma Lucy can not help.
✔ Grandma Lucy can not move.

❷ ✔ Kitty meows for help.
☐ Doggy meows for help.

❸ ✔ Everybody comes to help.
☐ Nobody comes to help.

42

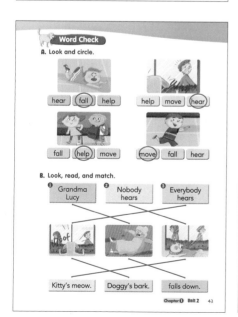

Word Check

A. Look and circle.

hear (fall) help   help move (hear)

fall (help) move   (move) fall hear

B. Look, read, and match.

❶ Grandma Lucy   ❷ Nobody hears   ❸ Everybody hears

Kitty's meow.   Doggy's bark.   falls down.

Chapter 1 Unit 2 43

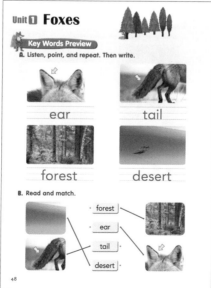

Unit 1 Foxes

Key Words Preview

A. Listen, point, and repeat. Then write.

ear

tail

forest

desert

B. Read and match.

forest
ear
tail
desert

48

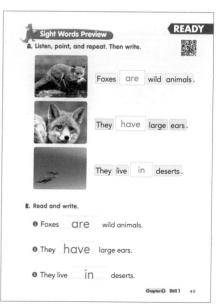

Sight Words Preview — READY

A. Listen, point, and repeat. Then write.

Foxes [are] wild [animals].

They [have] large [ears].

They [live] [in] [deserts].

B. Read and write.

❶ Foxes **are** wild animals.

❷ They **have** large ears.

❸ They live **in** deserts.

Chapter 2 Unit 1 49

Storybook 정답 59

Storybook 1 정답

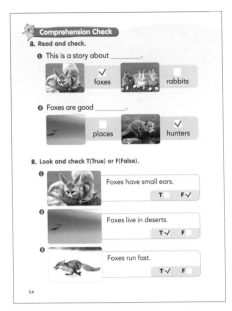

Comprehension Check

A. Read and check.

❶ This is a story about _____.
- foxes ✓
- rabbits

❷ Foxes are good _____.
- places
- hunters ✓

B. Look and check T(True) or F(False).

❶ Foxes have small ears.　T　F ✓

❷ Foxes live in deserts.　T ✓　F

❸ Foxes run fast.　T ✓　F

54

Word Check

A. Complete the puzzle.

```
        d
f o r e s t
        s   a
        e   i
e a r   t   l
```

B. Look, read, and match.

❶ They live　❷ They have　❸ They jump

a bushy tail.　high.　in forests.

Chapter 2　Unit 1　55

Unit 2　A Stupid Fox

Key Words Preview

A. Listen, point, and repeat. Then write.

little　catch

dinner　dirty

B. Read and circle.

catch / little
dinner / catch
dinner / (dirty)
dirty / little

56

Sight Words Preview　Part ❶　READY

A. Listen, point, and repeat. Then write.

They don't see the fox.

The fox catches one.

A dirty pig is not a tasty dinner.

B. Read and write.

❶ They don't see the fox.

❷ The fox catches one.

❸ A dirty pig is not a tasty dinner.

Chapter 2　Unit 2　57

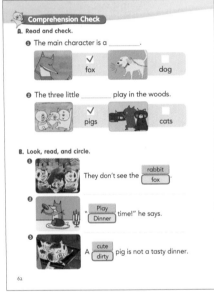

Comprehension Check

A. Read and check.

❶ The main character is a _____.
- fox ✓
- dog

❷ The three little _____ play in the woods.
- pigs ✓
- cats

B. Look, read, and circle.

❶ They don't see the rabbit / (fox)

❷ "Play / (Dinner) time!" he says.

❸ A cute / (dirty) pig is not a tasty dinner.

62

Word Check

A. Look and circle.

```
c k h d i n n e r
a z x c v b m n j
t q w l i t t l e
c p k j h g f d s
h v d i r t y b m
```

B. Look, read, and write.

Dinner time!

A dirty pig is not a tasty dinner.

Chapter 2　Unit 2　63

Unit 2　A Stupid Fox

Key Words Preview

A. Listen, point, and repeat. Then write.

want　tasty

wash　small

B. Read and match.

tasty
small
wash
want

64

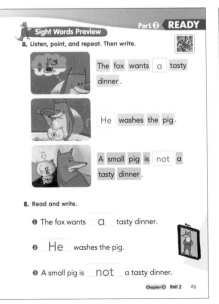

Sight Words Preview　Part ❷　READY

A. Listen, point, and repeat. Then write.

The fox wants a tasty dinner.

He washes the pig.

A small pig is not a tasty dinner.

B. Read and write.

❶ The fox wants a tasty dinner.

❷ He washes the pig.

❸ A small pig is not a tasty dinner.

Chapter 2　Unit 2　65

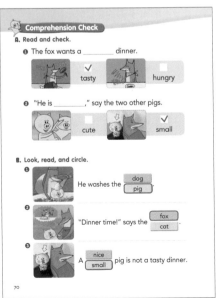

Comprehension Check

A. Read and check.

❶ The fox wants a _____ dinner.
- tasty ✓
- hungry

❷ "He is _____," say the two other pigs.
- cute
- small ✓

B. Look, read, and circle.

❶ He washes the dog / (pig)

❷ "Dinner time!" says the (fox) / cat

❸ A nice / (small) pig is not a tasty dinner.

70

60

Word Check

A. Look and circle.

- g f v d q [w a n t] z
- b p [t a s t y] v m x
- o f d z q c [w a s h]
- [s m a l l] v x p q w

B. Look, read, and write.

Dinner time!

A small pig is not a tasty dinner.

Chapter 2 Unit 2 71

Unit 2 A Stupid Fox

Key Words Preview

A. Listen, point, and repeat. Then write.

feed

food

tired

asleep

B. Read and circle.

- feed / (asleep)
- (tired) / feed
- (tired) / food
- asleep / (food)

72

Sight Words Preview

A. Listen, point, and repeat. Then write.

He feeds the pig more and more food!

Now the fox feels tired.

Bedtime, you stupid fox!

B. Read and write.

❶ He feeds the pig more and more food!

❷ Now the fox feels tired.

❸ Bedtime, you stupid fox!

Chapter 2 Unit 2 73

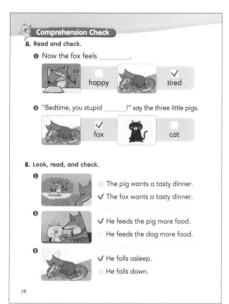

Comprehension Check

A. Read and check.

❶ Now the fox feels _____.

happy / ✓ tired

❷ "Bedtime, you stupid _____!" say the three little pigs.

✓ fox / cat

B. Look, read, and check.

❶
- ☐ The pig wants a tasty dinner.
- ✓ The fox wants a tasty dinner.

❷
- ✓ He feeds the pig more food.
- ☐ He feeds the dog more food.

❸
- ✓ He falls asleep.
- ☐ He falls down.

78

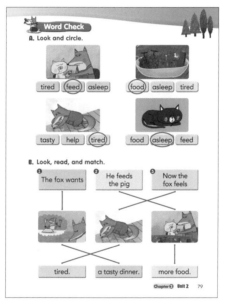

Word Check

A. Look and circle.

- tired / (feed) / asleep
- (food) / asleep / tired
- tasty / help / (tired)
- food / (asleep) / feed

B. Look, read, and match.

❶ The fox wants
❷ He feeds the pig
❸ Now the fox feels

tired. / a tasty dinner. / more food.

Chapter 2 Unit 2 79

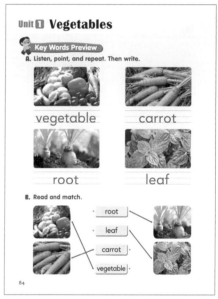

Unit 1 Vegetables

Key Words Preview

A. Listen, point, and repeat. Then write.

vegetable

carrot

root

leaf

B. Read and match.

- root
- leaf
- carrot
- vegetable

84

Sight Words Preview

A. Listen, point, and repeat. Then write.

We eat some plants.

A carrot is a vegetable.

It is a fruit.

B. Read and write.

❶ We eat some plants.

❷ A carrot is a vegetable.

❸ It is a fruit.

Chapter 3 Unit 1 85

Comprehension Check

A. Read and check.

❶ This is a story about _____.

✓ vegetables / animals

❷ A carrot is a _____.

☐ fruit / ✓ root

B. Look and check T(True) or F(False).

❶ We eat some plants.
T ✓ F

❷ A pumpkin is a root.
T F ✓

❸ Cabbage is a leaf.
T ✓ F

90

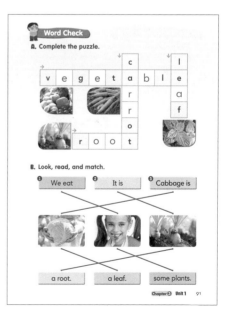

Word Check

A. Complete the puzzle.

						c		l
v	e	g	e	t	a	b	l	e
						r		a
						r		f
						o		
		r	o	o	t			

B. Look, read, and match.

❶ We eat
❷ It is
❸ Cabbage is

a root. / a leaf. / some plants.

Chapter 3 Unit 1 91

Storybook 1 정답

Unit 2 The Great Turnip

Key Words Preview

A. Listen, point, and repeat. Then write.

farmer turnip

wife pull

B. Read and circle.

pull / (wife) ... (farmer) / turnip

wife / (turnip) ... farmer / (pull)

92

Sight Words Preview — Part 1 READY

A. Listen, point, and repeat. Then write.

It's a great turnip.

The farmer calls his wife.

They pull the turnip together.

B. Read and write.

1 It's a great turnip.

2 The farmer calls his wife.

3 They pull the turnip together.

Chapter 3 Unit 2 93

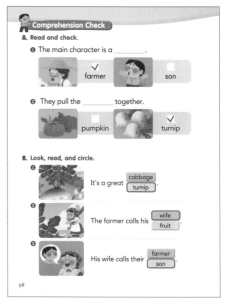

Comprehension Check

A. Read and check.

1 The main character is a _____.
✓ farmer   son

2 They pull the _____ together.
pumpkin   ✓ turnip

B. Look, read, and circle.

1 It's a great (turnip) [cabbage]

2 The farmer calls his (wife) [fruit]

3 His wife calls their (son) [farmer]

98

Word Check

A. Look and circle.

z	f	a	r	m	e	r	x	p
w	q	b	v	z	w	t	c	k
i	a	z	p	u	l	l	v	m
f	p	d	g	v	w	q	b	x
e	k	y	t	u	r	n	i	p

B. Look, read, and write.

It's a great turnip.

It's a big turnip.

It's a big, great turnip.

Chapter 3 Unit 2 99

Unit 2 The Great Turnip

Key Words Preview

A. Listen, point, and repeat. Then write.

king queen

like gold

B. Read and match.

like
king
queen
gold

100

Sight Words Preview — Part 2 READY

A. Listen, point, and repeat. Then write.

Let's take the turnip to the king.

The king will like it!

The king gives gold to the farmer.

B. Read and write.

1 Let's take the turnip to the king.

2 The king will like it!

3 The king gives gold to the farmer.

Chapter 3 Unit 2 101

Comprehension Check

A. Read and check.

1 "Let's take the _____ to the king," says the farmer.
✓ turnip   pumpkin

2 "The _____ will like it!" says his wife.
✓ king   son

B. Look, read, and circle.

1 "The (king) [queen] will like it!" says his son.

2 "I like it!" says the (king) [farmer].

3 The king gives (gold) [turnip] to the farmer.

106

Word Check

A. Look and circle.

k i n g q w a n z t

b q u e e n v m x

h g f w u q l i k e

s q w z g o l d b v

B. Look, read and write.

Let's take the turnip to the king.

The king will like it!

The queen will like it!

Chapter 3 Unit 2 107

Unit 2 The Great Turnip

Key Words Preview

A. Listen, point, and repeat. Then write.

rich man

give shout

B. Read and circle.

(man) / give ... give / (shout)

rich / (shout) ... (man) / rich

108

62

Sight Words Preview

A. Listen, point, and repeat. Then write.

Let's [take] gold to the king .

The king [and] the queen [will] like [it]!

[What] [will] [I] [give] [you]?

B. Read and write.

❶ Let's __take__ gold to the king.

❷ The king __and__ the queen will like it.

❸ __What__ will I give you?

Chapter ❸ Unit 2 109

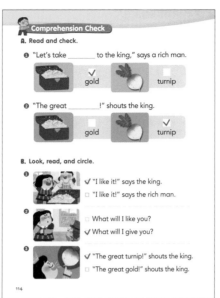

Comprehension Check

A. Read and check.

❶ "Let's take _____ to the king," says a rich man.

gold ☑   turnip ☐

❷ "The great _____!" shouts the king.

gold ☐   turnip ☑

B. Look, read, and circle.

❶ ☑ "I like it!" says the king.
☐ "I like it!" says the rich man.

❷ ☐ What will I like you?
☑ What will I give you?

❸ ☑ "The great turnip!" shouts the king.
☐ "The great gold!" shouts the king.

114

Word Check

A. Look and circle.

king (rich) shout

rich give (man)

shout (like) rich

man (give) like

B. Look, read, and match.

❶ Let's take gold
❷ The king and the queen will
❸ What will I

like it!
to the king.
give you?

Chapter ❸ Unit 2 115

Storybook 정답   63

Workbook 정답

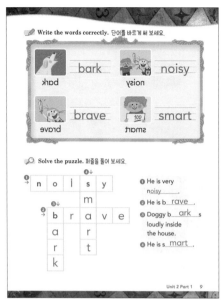

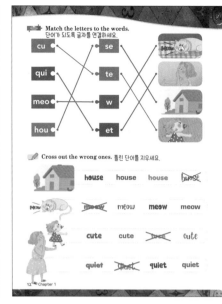

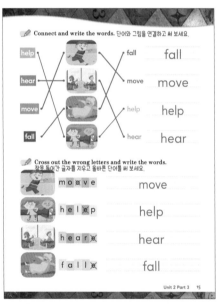

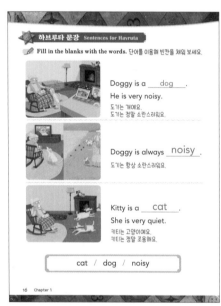

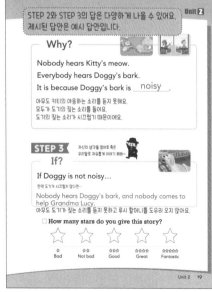

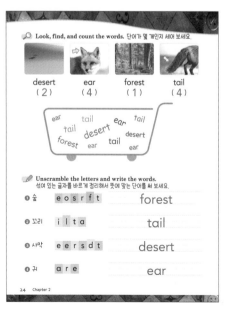

Write the words correctly. 단어를 바르게 써 보세요.

catch — catch
dinner — dinner
dirty — dirty
little — little

Solve the puzzle. 퍼즐을 풀어 보세요.

Crossword:
- little
- dinner
- dirty
- catch

① The three little pigs play in the woods.
② "He is dirty," say the two other pigs.
③ "Dinner time!" he says.
④ The fox catches one.

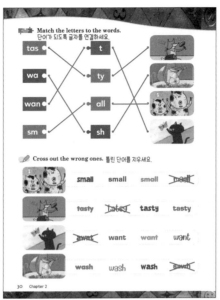

Match the letters to the words. 단어가 되도록 글자를 연결하세요.

tas • / wa • / wan • / sm •
• t / • ty / • all / • sh

Cross out the wrong ones. 틀린 단어를 지우세요.

small small small ~~smoll~~
tasty ~~tatsy~~ tasty tasty
~~awnt~~ want want want
wash wash wash ~~wahs~~

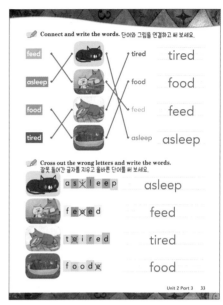

Connect and write the words. 단어와 그림을 연결하고 써 보세요.

feed — tired — tired
asleep — food — food
food — feed — feed
tired — asleep — asleep

Cross out the wrong letters and write the words. 잘못 들어간 글자를 지우고 올바른 단어를 써 보세요.

a s x l e e p — asleep
f e x e d — feed
t x i r e d — tired
f o o d x — food

하브루타 문장 Sentences for Havruta

Fill in the blanks with the words. 단어를 이용해 빈칸을 채워 보세요.

The three little pigs play in the woods.
The fox catches one.
아기 돼지 세 마리가 숲에서 놀아요.
여우가 그 중 한 마리를 잡아요.

The pig is dirty.
그 돼지는 더러워요.

The fox washes the pig.
여우는 돼지를 씻겨요.

dirty / catch / wash

OO에게 들려주고 싶은 문장이 있나요?
그 문장을 외워서 OO 앞에서 큰 소리로 말해 보세요.

The pig is small.
그 돼지는 작아요.

The fox feeds the pig.
여우는 돼지를 먹여요.

The fox feels tired and falls asleep.
여우는 피곤해서 잠들어 버려요.

asleep / feed / tired / small

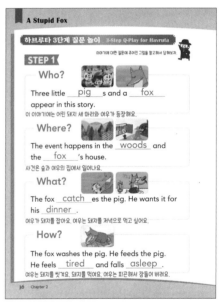

A Stupid Fox

하브루타 3단계 질문 놀이 3-Step Q-Play for Havruta
이야기에 대한 질문에 주어진 그림을 참고해서 답해보자

STEP 1

Who?
Three little pigs and a fox appear in this story.
이 이야기에는 어린 돼지 세 마리와 여우가 등장해요.

Where?
The event happens in the woods and the fox's house.
사건은 숲과 여우의 집에서 일어나요.

What?
The fox catches the pig. He wants it for his dinner.
여우가 돼지를 잡아요. 여우는 돼지를 저녁으로 먹고 싶어요.

How?
The fox washes the pig. He feeds the pig. He feels tired and falls asleep.
여우는 돼지를 씻겨요. 돼지를 먹여요. 여우는 피곤해서 잠들어 버려요.

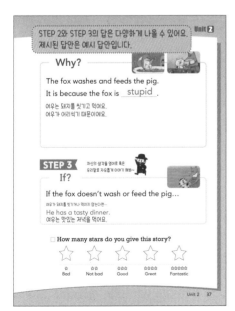

STEP 2와 STEP 3의 답은 다양하게 나올 수 있어요.
제시된 답안은 예시 답안입니다.

Why?
The fox washes and feeds the pig.
It is because the fox is stupid.
여우는 돼지를 씻기고 먹여요.
여우가 어리석기 때문이에요.

STEP 3
자신의 생각을 영어로 혹은 우리말로 자음 색깔 펜으로 이어서 써봐~

If?
If the fox doesn't wash or feed the pig...
여우가 돼지를 씻기거나 먹이지 않는다면~
He has a tasty dinner.
여우는 맛있는 저녁을 먹어요.

☐ How many stars do you give this story?
★ ★ ★ ★ ★
Bad / Not bad / Good / Great / Fantastic

Look, find, and count the words. 단어가 몇 개인지 세어 보세요.

vegetable (2)
carrot (4)
leaf (3)
root (3)

root carrot vegetable leaf vegetable carrot root leaf carrot carrot leaf root

Unscramble the letters and write the words.
섞여 있는 글자를 바르게 정리해서 뜻에 맞는 단어를 써 보세요.

① (나뭇)잎 a e f l — leaf
② 당근 r r o a c t — carrot
③ 뿌리 o t r o — root
④ 채소, 야채 a e b e l e v t g — vegetable

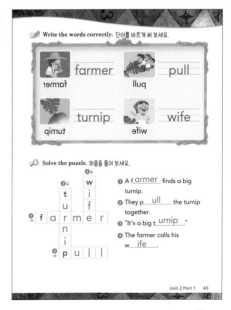

Write the words correctly. 단어를 바르게 써 보세요.

farmer — farmer
pull — pull
turnip — turnip
wife — wife

Solve the puzzle. 퍼즐을 풀어 보세요.

Crossword:
- turnip
- wife
- farmer
- pull

① A farmer finds a big turnip.
② They pull the turnip together.
③ "It's a big turnip."
④ The farmer calls his wife.

Workbook 정답

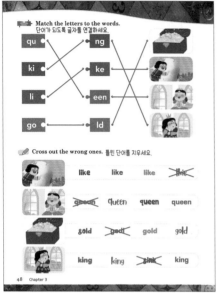

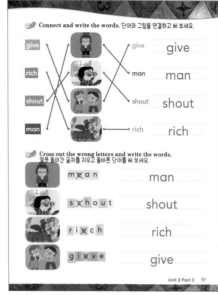

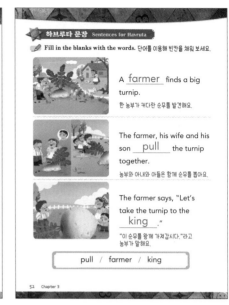

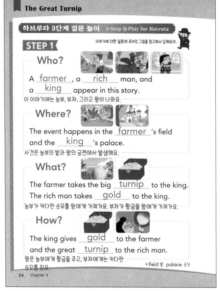

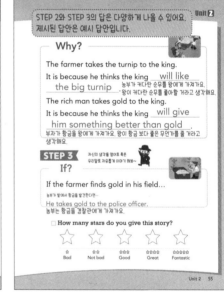

본문 해석

Chapter 1 Cats and Dogs
Unit 1 Cats and Dogs

Cats and dogs are good friends
Cats like to play.
Dogs like to walk.
Cats like to sleep.
Dogs like to fetch.
Cats like to hunt.
Dogs like to smell.
Which do you like? (37)

고양이와 개

고양이와 개는 좋은 친구예요.
고양이는 장난치는 걸 좋아해요.
개는 산책하는 걸 좋아해요.
고양이는 잠자는 걸 좋아해요.
개는 공 물어 오기를 좋아해요.
고양이는 사냥을 좋아해요.
개는 냄새 맡는 걸 좋아해요.
여러분은 고양이와 개 중 어떤 게 좋은가요?

Unit 2 Doggy the Noisy Dog

Doggy is a dog.
He is smart.
He is brave.
But he is very noisy.
Doggy barks loudly inside the house.
He barks loudly outside the house.
"What a noisy dog, Doggy!" says Grandma Lucy.
Kitty is a cat.
She is smart.
She is cute.
And she is very quiet.
Kitty meows quietly inside the house.
She meows quietly outside the house.
"What a nice cat, Kitty!" says Grandma Lucy.
One day, Grandma Lucy falls down.
She can not move.
Kitty meows for help.
Nobody hears Kitty's meow.
Doggy barks for help.
Everybody hears Doggy's bark.
Everybody comes to help.
"Good boy, Doggy!" they say. (109)

떠들썩한 강아지, 도기

도기는 개예요.
도기는 영리해요.
도기는 용감해요.
근데 도기는 정말 소란스러워요.
도기는 집 안에서 우렁차게 짖어요.
도기는 집 밖에서도 우렁차게 짖어요.
"도기, 정말 시끄럽구나!"하고 루시 할머니가 말해요.
키티는 고양이예요.
키티는 영리해요.
키티는 귀여워요.
그리고 키티는 정말 조용하답니다.
키티는 집 안에서 조용히 야옹 해요.
키티는 집 밖에서도 조용히 야옹 해요.
"아이고 착해라, 키티!"하고 루시 할머니는 말해요.
어느 날, 루시 할머니가 쓰러져요.
할머니는 움직일 수가 없어요.
키티가 도와 달라고 야옹 해요.
아무도 키티 소리를 듣지 못해요.
도기가 도와 달라고 짖어요.
모두가 도기 소리를 들어요.
모두 도와주러 와요.
"잘했어, 도기!"라고 사람들이 말해요.

Chapter 2 Foxes
Unit 1 Foxes

Foxes are wild animals.
They have a bushy tail.
They have large ears.
Foxes live in many places.
They live in forests.
They live in deserts.
Foxes are good hunters.
They run fast.
They jump high. (37)

여우

여우는 야생 동물이에요.
털이 복슬복슬한 꼬리를 가지고 있어요.
귀가 커요.
여우는 다양한 곳에서 살아요.
숲에서도 살고
사막에서도 살아요.
여우는 사냥을 잘해요.
달리기도 빠르고
높이 점프도 해요.

Unit 2 A Stupid Fox

The Three little pigs play in the woods.
They don't see the fox.
The fox catches one.
"Dinner time!" he says.
"He is dirty," say the two other pigs.
"A dirty pig is not a tasty dinner."
The fox wants a tasty dinner.
He washes the pig.
"Dinner time!" says the fox.
"He is small," say the two other pigs.
"A small pig is not a tasty dinner."
The fox wants a tasty dinner.
He feeds the pig more and more food.
Now the fox feels tired.
He falls asleep.
"Bedtime, you stupid fox!" say the three little
pigs. (102)

Chapter 3 Vegetables
Unit 1 Vegetables

We eat some plants.
They are vegetables.
A carrot is a vegetable. It is a root.
A pumpkin is a vegetable. It is a fruit.
Cabbage is a vegetable. It is a leaf.
Which do you like to eat? (40)

Unit 2 The Great Turnip

A farmer finds a big turnip.
"It's a big turnip."
The farmer calls his wife.
"It's a great turnip."
His wife calls their son.
"It's a big, great turnip."
They pull the turnip together.
"Let's take the turnip to the king," says the farmer.
"The king will like it!" says his wife.
"The queen will like it!" says his son.
"I like it!" says the king.
The king gives gold to the farmer.
"Let's take gold to the king," says a rich man.
"The king and the queen will like it!"
"What will they give us? says his wife.
"I like it!" says the king.
"What will I give you?"
"The great turnip!" shouts the king. (115)

멍청한 여우

아기 돼지 세 마리가 숲에서 놀고 있어요.
돼지들은 여우를 보지 못해요.
여우가 돼지 한 마리를 잡아요.
"저녁 식사로 딱이네!"라고 여우가 말해요.
"그 애는 더러워." 다른 돼지 두 마리가 말해요.
"더러운 돼지가 맛있는 저녁 식사는 아니지."
여우는 맛있는 저녁을 먹고 싶어요.
여우가 돼지를 씻겨요.
"이제 저녁 먹을 시간!" 하고 여우가 말해요.
"그 애는 작아." 다른 돼지 두 마리가 말해요.
"작은 돼지가 맛있는 저녁 식사는 아니지."
여우는 맛있는 저녁을 먹고 싶어요.
여우가 돼지에게 점점 더 많은 음식을 먹여요.
이제 여우는 피곤해요.
여우가 잠들어요.
"잘 시간이다, 이 멍청한 여우야!" 아기 돼지 세 마리가 말해요.

채소

어떤 식물은 우리가 먹어요.
바로 채소가 그 식물들이에요.
당근은 채소예요. 당근은 뿌리에 해당해요.
호박도 채소예요. 호박은 열매예요.
양배추도 채소예요. 양배추는 잎이에요.
여러분은 어떤 채소를 좋아하나요?

커다란 순무

한 농부가 커다란 순무를 발견해요.
"커다란 순무구만."
농부는 아내를 불러요.
"아주 커다란 순무네요."
농부의 아내가 아들을 불러요.
"엄청 커다란 순무네요."
농부와 아내와 아들은 함께 순무를 뽑아요.
"이 순무를 왕께 가져갑시다."라고 농부가 말해요.
"왕이 좋아하실 거예요!" 농부의 아내가 말해요.
"왕비도 좋아하실 거예요!" 아들이 말해요.
"마음에 드는구나!"라고 왕이 말해요.
왕은 농부에게 황금을 줘요.
"황금을 왕께 가져갑시다." 라고 부자가 말해요.
"왕과 왕비가 좋아하실 거예요!"
"왕과 왕비가 우리에게는 뭘 줄까요?" 부자의 아내가 말해요.
"마음에 드는구나!"라고 왕이 말해요.
"그대에게 무엇을 줄까?"
"아주 커다란 순무를 주겠네!" 왕이 소리쳐요.

진짜진짜

초등
영어 읽기

하브루타 워크북
1